EcoDeep-SIP Workshop II

Dialogue on practical utility of EIA technology - Proponent, potential users, administrator and independent organization

14-15 March 2017
JAMSTEC Tokyo Office,
Uchisaiwaicho, Tokyo, Japan

ISA Technical Study No: 18

ISA TECHNICAL STUDY SERIES

Technical Study No. 2   Polymetallic Massive Sulphides and Cobalt-Rich Ferromanganese Crusts: Status and Prospects

Technical Study No. 3   Biodiversity, Species Ranges and Gene Flow in the Abyssal Pacific Nodule Province: Predicting and Managing the Impacts of Deep Seabed Mining

Technical Study No. 4   Issues associated with the Implementation of Article 82 of the United Nations Convention on the Law of the Sea

Technical Study No. 5   Non-Living Resources of the Continental Shelf Beyond 200 Nautical Miles: Speculations on the Implementation of Article 82 of the United Nations Convention on the Law of the Sea

Technical Study No. 6   A Geological Model of Polymetallic Nodule Deposits in the Clarion-Clipperton Fracture Zone

Technical Study No. 7   Marine Benthic Nematode Molecular Protocol Handbook (Nematode Barcoding)

Technical Study No. 8   Fauna of Cobalt-Rich Ferromanganese Crust Seamounts

Technical Study No. 9   Environmental Management of Deep-Sea Chemosynthetic Ecosystems: Justification of and Considerations for a Spatially-Based Approach

Technical Study No. 10   Environmental Management Needs for Exploration and Exploitation of Deep Sea Minerals

Technical Study No. 11   Towards the Development of a Regulatory Framework for Polymetallic Nodule Exploitation in the Area.

Technical Study No. 12   Implementation of Article 82 of the United Nations Convention on the Law of the Sea

Technical Study No. 13   Deep Sea Macrofauna of the Clarion-Clipperton Zone

Technical Study No. 14   Submarine Cables and Deep Seabed Mining

Technical Study No. 15   A Study of Key terms in Article 82 of the United Nations Convention on the Law of the Sea

Technical Study No. 16   Environmental Assessment and Management for Exploitation of Minerals in the Area

Technical Study No. 17   Towards an ISA Environmental Management Strategy for the Area

EcoDeep-SIP Workshop II

DIALOGUE ON PRACTICAL UTILITY OF EIA TECHNOLOGY - PROPONENT, POTENTIAL USERS, ADMINISTRATOR AND INDEPENDENT ORGANIZATION
Fact Findings and Recommendations

Technical Study No. 18

Tokyo, 14-15 March 2017
JAMSTEC Tokyo Office, Uchisaiwaicho, Tokyo, Japan

Supervising Editor: International Seabed Authority (ISA)
                       Editors: JAMSTEC & IFREMER
Contributing Writers: Tomohiko Fukushima (JAMSTEC)
                       Aki Tanaka (JAMSTEC)
                       Yves Henocque (IFREMER)

NATIONAL LIBRARY OF JAMAICA CATALOGUING-IN-PUBLICATION DATA

International Seabed Authority
    EcoDeep SIP Workshop II : dialogue on practical utility of EIA technology :
Proponent, potential users, administrators and independent organization, 14-15 March 2017, JAMSTEC Tokyo Office, Uchisaiwaicho, Tokyo, Japan.

        p. : ill. , maps; cm – (ISA technical study; no. 18)

ISBN 978-976-8241-50-4 (pbk)
ISBN 978-976-8241-51-1 (ebk)

1.   Ocean mining – Environmental aspects
2.   Marine mineral resources – Environmental aspects
I .   Title   II. Series

333.9164   dc 23

Contents

Session 1

DAY 2: 15 MARCH 2017

Session 2

Session 3

ACRONYMS

ABNJ	Areas beyond national jurisdiction
ECODEEP	Ecological Aspects of Hydrothermal Vents and Massive Sulphide Deposits
EBSA	Ecologically and Biologically Significant Areas
EIA	Environmental Impact Assessment
IFREMER	Institut Français de Recherche pour l'Exploitation de la Mer
IRD	Institut de Recherche pour le Développement
ISA	International Seabed Authority
JAMSTEC	Japan Agency for Marine-Earth Science and Technology
NIES	National Institute for Environmental Studies
OBIS	Ocean Biogeographic Information System
SIP	Strategic Innovation Promotion Program
SPC	Secretariat of the Pacific Community
TUMSAT	Tokyo University of Marine Science and Technology
UNCLOS	United Nations Convention on the Law of the Sea
YNU-DEEPS	Yokohama National University, Deep-sea Resource Exploration and Environment Protection Study

INTRODUCTION

Hosts and Cooperative Organizations

Japan Agency for Marine-Earth Science and Technology (JAMSTEC), National Institute of Environmental Studies (NIES), Yokohama National University Deep-sea Resource Exploration and Environmental Protection Study (YNU-DEEPS), Tokyo University of Marine Science and Technology (TUMSAT), Institut français de recherche pour l'exploitation de la mer (IFREMER)

Invited Participants

International Seabed Authority (ISA)

>**Michael Lodge** (Secretary-General)
>**Alfonso Ascencio-Herrera** (Legal Counsel and Deputy to the Secretary-General)
>**Sandor Mulsow** (Director of the Office of Environmental Management and Mineral Resources)

Secretariat of the Pacific Community (SPC)

>**Akuila Tawake** (Head of Geo-surveys and Geo-resources Sector)

Institut de Recherche pour le Développement (IRD), Nouvelle-Calédonie

>**Pierre-Yves Le Meur** (Research Director)

Institut Français de Recherche pour L'exploitation de la Mer (IFREMER)

>**Yves Henocque** (International Relations Officer and Maritime Policy and Governance Senior Adviser)
>**Jean-Marc Daniel** (Director, Department of Physical Resources and Deep-Sea Ecosystems)

Backgrounds and Objectives

Appropriate environmental impact assessment technologies (EIA related-technologies) are required for development of deep-sea mineral resources. However, the word "appropriate" includes various meanings. From the standpoint of ISA contractors, needless to say that reliability and economic rationality are important concerns. On the other hand, as resources in areas beyond national jurisdiction (ABNJ) are 'common heritage of mankind', costly methods and technics that can be only applied by developed countries do not meet the United Nations Convention on the Law of the Sea (UNCLOS) spirit and requirement. Considering the international trend towards raising the level of environmental protection, EIA technologies should guarantee accuracy and precision and be cost-effective. Furthermore, the concept and spirit of the common heritage of mankind require the development of activities to be made for the benefit of all countries and communities. In such circumstances, and in order to determine what could be the components of an ideal EIA, dialogue between users, administrators and technical providers is needed, with the help of a third party acting as a referee.

The Japanese government has initiated a program devoted to the development of EIA related-technologies as required by the international society. As part of this project, during the summer of 2014, the Japan Agency for Marine-Earth Science and Technology (JAMSTEC) and the National Institute of Environmental Studies (NIES), jointly with the Institut français de recherche pour l'exploitation de la mer (IFREMER), convened an international workshop at the French Embassy named EcoDEEP-SIP workshop. During this workshop, including an ISA representative, needs concerning both procedures and technologies for environmental research leading to the development of deep-sea mineral resources were unanimously confirmed.

In this context, to follow up on procedures and technologies that are under consideration, a second EcoDEEP-SIP workshop was held on 14-15 March 2017 in Tokyo. Basically, this second workshop followed the previous framework including representatives from SIP, IFREMER, SPC (Secretariat of the Pacific Community), IRD (French Polynesia), and ISA. During this workshop, the Japan SIP national program played the role of the technical knowledge provider, SPC (Geosciences Division) of the potential user, ISA of the administrator, and IFREMER was acting as the impartial third party and referee.

Two Initiatives at Stake

1. Ecological Aspects of Hydrothermal Vents and Massive Sulfide Deposits (EcoDeep)

EcoDeep is a 4-year project that was launched between IFREMER and JAMSTEC regarding the assessment of environmental impacts of deep seabed mining, addressing both scientific issues and technological challenges and more generally contributing to building up a strategy for ecosystem-based management of future deep seabed mining activities. Acquisition of fundamental knowledge on the structure, dynamic and function of deep sea ecosystems is of tremendous importance to assess and predict the spatio-temporal scales of environmental impacts due to mineral mining as well as their consequences on ecosystem functioning and services.

Contractors of exploration or exploitation licenses for deep-sea mineral resources are bind by national or international regulations to perform an environmental impact assessment of their activities. Such assessment in the deep sea needs to be both robust and cost-effective requiring the design of environmental impact assessment studies protocol and long-term monitoring, with methods applicable to developing country's technology. Both JAMSTEC and IFREMER, in bilateral and multilateral ways, are thus collaborating scientifically and technologically to contribute to the development of environmental guidelines as compiled by the International Seabed Authority (ISA).

**A collaborative ecological approach towards
ecosystem-based management of deep seabed mining**

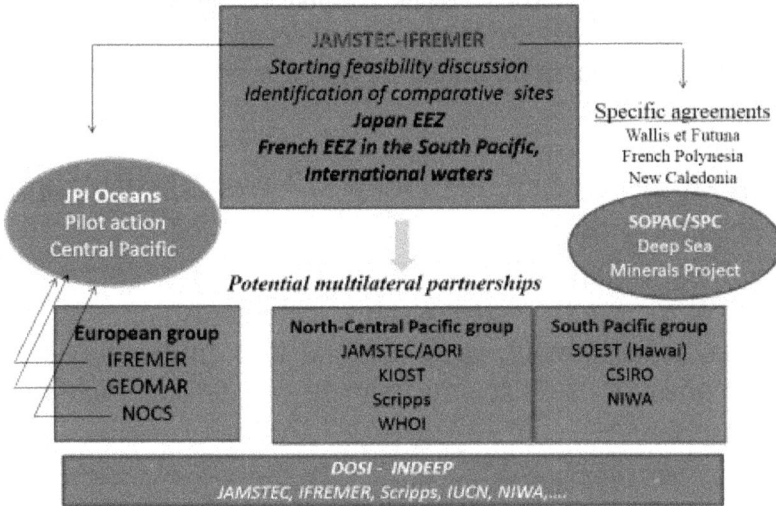

2. Strategic Innovation Promotion Program (SIP)

The Strategic Innovation Promotion Program (SIP) is a Japanese cross-ministerial interdisciplinary program among government agencies, academic institutes and private sector addressing eleven strategic themes. One of them is the 'Next-generation technology for ocean resources exploration', more particularly deep mineral resources.

It is clearly oriented towards, 1) the launching of commercial-level exploration of ocean resources and, 2) the establishment of global standards.

Towards such a goal, scientific research on formation processes, technology development, ecological surveys and deep sea observing systems are needed.

The second EcoDeep-SIP workshop was the opportunity to prepare and discuss a first set of four 'SIP Protocol Series' covering, 1) the

'Application of environmental metagenomic analysis for environmental impact assessments', 2) the 'Genetic connectivity survey manuals', 3) 'A rapid method to analyze meiofaunal assemblages using an imaging flow cytometer', and 4) the 'Acquisition of long-term monitoring images near the deep seafloor by Edokko Mark I'.

Recalling the first EcoDeep-SIP workshop's main recommendations

- **We need deep-ocean stewardship as a transdisciplinary, multi-sectoral and multi-stakeholder endeavor within and beyond national jurisdictions** *provided:*
    - We are addressing huge areas with connected though highly heterogeneous living communities, which are remote and far from our bases, whilst we are financially and technology limited, leading to lack of needed scientific knowledge in regard to biodiversity response and ecosystem services recovery;
    - Vast area and long residence times typify deep sea environments, meaning that even fast processes on small spatial scales (e.g. hydrothermal vents) create massive services, although in many cases the processes are far removed from their resultant services (mainly interrelated regulating and provisioning services);
    - While there is a patent lack of public awareness regarding the richness/vulnerability of the deep ocean, we are working under different governance regimes covering the international waters and countries' EEZ.

- **We know that the deep seabed may be a very complex place, interconnected on a vast range of scales,** *more particularly regarding the SMS sites:*
    - Depending on taxon, connectivity can occur over very large areas leading to large-scale baseline studies being needed of community and trophic structures and exchange

of genes within a single basin whilst genetic diversity seem specific to each basin;

o Hence the importance of getting oceanographic data more particularly in regard to circulation in basins.

- **From bottom to surface, the most significant impacts expected from deep-sea mining are:**

  o Mineral extraction, cutting and return plumes (particles and toxicity), extraction machines and riser noise during regular activities whilst spill over may occur from incidents and ship-related activities;

  o Hence mitigation priorities are: protection of areas from damage or impacts, e.g. representative MPAs (including APEI, Preservation Reference Zone, and other zoning options), limit spread of return plume, limit generation of plumes by mining tools, limit noise intensity and frequency, slow start to mining, spacing and timing of mining, along one overall rule where mitigation approaches assume appropriate monitoring and adaptive management (use of technologies and knowledge as become available) based on systematically minimizing mining impact;

  o Cumulative impacts might be looked at with respect to multiple mines in the same zone (e.g. CCZ), multiple industry activities in the same area (e.g. mining and fisheries), and in the context of increasing climate-stressors (warming, acidification, de-oxygenation) and eutrophication events.

- **Ways to reflect those priorities in the environmental assessment process:**

  o Minimize mining impacts: high quality Regional Environmental Assessment, Environmental Impact Assessments and Ecological Risk Assessment are all important steps in the regulatory process that require a strong partnership between research and industry;

  o Make measurement/sampling activities as minimally intrusive as possible: technology development is needed for exploration equipment, production equipment (including

noise reduction) and monitoring equipment;
- o While adapted, standardized and commonly agreed monitoring devices and procedures must be developed jointly with require regulation/standards to be set up under ISA.

This last recommendation is at the core of the EcoDeep-SIP second workshop.

Programme

Day 1: 14 March 2017

9:00	Welcome Address **Tetsuro Urabe**, SIP Program Director, Professor Emeritus of the University of Tokyo
9:15	Introduction of Invited Participants **Tomohiko Fukushima**, Assistant Director, JAMSTEC
9:30	Keynote Speech "Role of Science for Conservation of Marine Biodiversity" **Yoshihisa Shirayama**, Executive Director, JAMSTEC
10:00	Keynote Speech "Current topics of ISA's efforts concerning Ocean Environmental Conservation" **Michael W. Lodge**, Secretary-General, ISA
10:30	Coffee break
10:50	Deep Seabed Minerals Development in the Pacific Islands Region – Current Status and Challenges **Akuila Tawake**, Head of Geo-surveys and Geo-resources Sector, SPC
11:20	Mining and environmental impact assessment in French territory in the Pacific **Pierre-Yves Le Meur**, Research Director, Institute of Research and Development
11:50	EU project for Environmental Impact Assessment in EU **Jean-Marc Daniel,** Director of Department of Physical Resources and Deep-Sea Ecosystems, IFREMER
12:20	Lunch break

Session 1	Chair: **Yves Henocque** (IFREMER) Vice-Chair: **Tomohiko Fukushima** (JAMSTEC) Rapporteur: **Dhugal J. Lindsay** (JAMSTEC) and **Makoto Seta** (Yokohama City University)
13:40 13:50 14:20	*1-1 Objectivity and Reliability of data* Brief orientation : **Tomohiko Fukushima**, JAMSTEC Introduction of Metagenomics : **Miyuki Nishijima**, JAMSTEC Discussion
14:40 14:50 15:20	*1-2 Working Efficiency* Brief orientation : **Tomohiko Fukushima**, JAMSTEC A rapid method to analyze meiofaunal assemblages using an Imaging Flow Cytometer : **Tomo Kitahashi**, JAMSTEC Discussion
15:40	Coffee break and Poster session
16:20 16:30 17:00	*1-3 Technical Feasibility* Brief orientation : **Tomohiko Fukushima**, JAMSTEC Introduction of the Edokko Mark I : **Tetsuya Miwa**, JAMSTEC Discussion
17:20	End of Day 1
18:00	Reception at JAMSTEC Tokyo Office

Day 2: 15 March 2017

Session 2	Chair : **Jean-Marc Daniel**, IFREMER Vice-Chair: **Masanobu Kawachi**, NIES Rapporteur: **Dhugal J. Lindsay**, JAMSTEC and **Makoto Seta**, Yokohama City University
9:00 9:10 9:40	*2-1 Unexpected Impact* Brief orientation: **Masanobu Kawachi**, NIES Environmental pollution monitoring system using bioassay: **Hiroshi Koshikawa**, NIES Discussion
10:00 10:10 10:40	*2-2 Potential technology* Brief orientation: **Hiroyuki Yamamoto**, JAMSTEC Turbulence measurement for plume diffusion estimation: **Yasuo Furushima**, JAMSTEC Discussion
11:00	Break
Session 3	Chair: **Yves Henocque** and Jean-Marc Daniel, IFREMER Vice-Chair: **Tomohiko Fukushima**, JAMSTEC and **Masanobu Kawachi**, NIES
11:10 12:40	*3-1 Wrap up and Recommendation* Rapporteur's report and Discussion Closing address : **Tetsuro Urabe,** SIP Program Director
12:50	End of the Workshop

Keynote Addresses & Regional Reports

DAY1: 14 MARCH 2017

Key Note Address: The Scientist's Word

Knowledge and conservation of marine biodiversity and the World Ocean Census legacy
Yoshihisa Shirayama, JAMSTEC Executive Director for Science

Launched in 2000, the decade-long Census of Marine Life, gathering more than 2,700 scientists from more than 80 nations, spanned all ocean realms, from coast to abyss, from the North Pole to Antarctic shores, from the long past to the future to recognize the 'known, knowable, and unknowable', from whales to microbes. It systemically compiled the existing known and new discoveries into the Ocean Biogeographic Information System (OBIS – http://iOBIS.org), which is now incorporated into UNESCO Intergovernmental Oceanographic Commission (IOC) as part of the International Oceanographic Data and Information Exchange (IODE) programme. Using conventional research ships and sampling, divers and submersible vehicles, genetic identification, electronic and acoustic tagging, listening posts and communicating satellites, it described hundreds of new species, proving the existence of numerous knowable in the ocean.

The Census of marine life generated discoveries on the diversity and distribution of microbes, millions of them in any cubic cm of sea water, the largest source of marine biomass but, at the current technological status, still largely unknowable.

Regarding the 'known' biodiversity, OBIS, as a publically accessible data portal, played a pivot role, delivering up the results of over 800 existing, quality controlled data collections, including all the data gathered by Census projects. For example, Criterion 6 concerning biological diversity defines an EBSA (Ecologically and biologically significant areas) as an area containing relatively more diversity of ecosystems, habitats, communities, or species, or an area with more genetic diversity. Thanks to OBIS, the Convention of Biological Diversity (CBD) has been provided more than 22 million records, with estimated biodiversity indices.

But, in spite of these efforts, effective and efficient observation of more than 200,000 species of marine animals and perhaps tens of millions of types of marine microbes have still to be achieved, more especially in the deep sea. Existing long-time series of marine life are generally rare and even rarer in the deep sea. This paucity is in contrast with the more numerous marine chemical and physical data series captured by remote sensing, drifting buoys, or active float systems.

Whatsoever, as a precautionary principle, it is of the utmost importance to use the Census existing data regarding the abyssal plains or the seamounts, in the design of "Preservation Reference Area" network like it is the case in the Clarion-Clipperton zone of the central Pacific Ocean as managed by the International Seabed Authority (ISA). Another example goes with the South Pacific Regional Fisheries Management Organization that have used indicator species to predict where habitats sensitive to fishing might occur in data poor regions.

As encouraged by ISA, it is therefore the role of science to use the existing data and to produce new ones in order to answer the question "to do or not to do" in regard to possible future deep sea mining activities. This must also be done along 'fair' procedures, i.e. procedures that can then be used by anybody for environmental impact assessment or re-assessment when needed.

Yet, there will always be the 'unknowable', meaning the uncertainty surrounding potential threats to the environment, to manage with. In response, the precautionary approach provides a fundamental policy basis to anticipate, avoid and mitigate threats to the environment. It is therefore not about imposing prohibition but about being proactive calling on users and decision-makers to place the powers of scientific inquiry, technological innovation, political decision-making, legislative enactment, economic production and personal skills in the service of new and creative ways of behaving and doing.

Taking into consideration the cultural and place specificities (development value for human), the stake of science is to develop a knowledge system that allows as far as possible the assessment of nature value (loss due to development) through the measurement of ecosystem services and the use of an adaptive management approach, i.e., where any new information gained through monitoring and further research or information-gathering can then be fed back to inform further management and decision-making. While in some cases this may lead to the precautionary measure no longer being needed, in others it may lead to the determination that the threat is more serious than expected and that more stringent measures are required.

Key Note Address: The Administrator's Word

Current efforts by the International Seabed Authority (ISA) relating to environmental regulation of activities in the Area
M. Lodge, ISA Secretary-General

Legal mandate for environmental regulation
ISA was established 23 years ago, in 1994. However, its history goes back almost 50 years to the beginning of the discussions that led to the negotiation of a comprehensive United Nations Convention on the Law of the Sea (UNCLOS).

Environmental regulation is actually one of the most important tasks of ISA: UNCLOS Article 145 requires the Authority to adopt appropriate rules, regulations and procedures for, a) the prevention, reduction and control of pollution and other hazards to the marine environment, including the coastline, and b) the protection and conservation of the natural resources of the Area and the prevention of damage to the flora and fauna of the marine environment.

This is a very precisely worded provision, which complements the general provisions of the Convention in relation to the protection of the marine environment contained in Part XII of the Convention. Those provisions in turn constitute the basic framework for the legal regime that establishes the obligations, powers and responsibilities of States with respect to the marine environment.

Article 192 establishes the overarching obligation of all States to protect and preserve the marine environment. Articles 194, 204 and 206 go on to describe the specific measures to be taken by States to prevent, reduce and control marine pollution as well as to ensure that activities under their jurisdiction or control do not cause pollution damage to other States and their environment, and that pollution does not spread beyond the areas where they exercise sovereign rights under the Convention.

In relation to the Area, Article 209 states that 'international rules, regulations and procedures shall be established in accordance with Part XI to prevent, reduce and control pollution of the marine environment from activities in the Area'. This provision therefore forms a direct link to Article 145.

Mentioning these legal provisions emphasize the fact that, as an institution created by the Convention, the Authority is an international organization with precisely defined and limited powers and functions. Any implied powers the Authority may have are expressly limited to those that are implicit and necessary for the exercise of its powers and functions with respect to activities in the Area.

In particular, definitions are important. The Authority's mandate is limited to 'activities in the Area', which are defined as exploration for and exploitation of deep seabed mineral resources. The Authority's main responsibility with regard to the marine environment is to 'prevent, reduce and control pollution and other hazards' to the marine environment, where 'pollution' is a defined term in Article 1 of the Convention.

To put it in more straightforward terms, the Authority's task is to set the conditions under which deep sea mining can proceed without causing serious harm to the marine environment. That means preventing, reducing and controlling known significant harmful effects as far as possible through regulatory mechanisms that require appropriate risk assessment, provide for long-term monitoring and management of environmental impacts and incentivizing engineering and mining planning solutions that minimize environmental damage.

For the purposes of the Authority's regulatory regime, deep sea mining is divided in two phases: exploration and exploitation. Whilst there may be some overlap between these phases in terms of activities that may be permitted, the primary objectives of the exploration phase are to identify mineable areas, carry out tests of equipment and conduct environmental baseline studies. The Authority has adopted regulations governing exploration for all three types of mineral resources –nodules, sulphides, and crusts- and is currently developing an exploitation code. Together these will form a complete mining code.

Environmental regulation during the exploration phase

The Exploration Regulations attempt to strike a balance between a precautionary approach to activities in the Area and an incremental approach to regulation, with an emphasis on gathering sufficient data during the early phase of exploration in order to determine the range of potential environmental impacts. This is a logical approach to take because the majority of activities carried out in the exploration phase have little or no detrimental effect on the marine environment.

The Regulations impose a duty on each contractor to 'take necessary measures to prevent, reduce and control pollution and other hazards to the marine environment arising from its activities in the Area as far as reasonably possible, applying a precautionary approach and best environmental practices'.

To give effect to this general duty, exploration contracts require contractors to submit an assessment of the potential environmental impacts of their proposed activities. Thereafter contractors are required to gather environmental baseline data as exploration activities progress and to establish environmental baselines. Contractors are also required to carry out monitoring programmes but also may be required to cooperate with the Authority and sponsoring States in the establishment and implementation of such monitoring programmes. In practice this means that contractors are required to submit an annual report on the implementation and results of their environmental monitoring programmes, including relevant data and information. These reports are then forwarded to the Legal and Technical Commission of the Authority which reviews them and makes such comments and recommendations as may be necessary to the Secretary-General, who then draws any relevant issues to the attention of contractors.

At a more practical level, the Regulations are supplemented by detailed 'Recommendations' for the guidance of contractors issued by the legal and Technical Commission. The first set of environmental recommendations was issued in 2001 and dealt with the assessment of possible environmental impacts arising from exploration for polymetallic nodules. The recommendations described the procedures to be followed in the acquisition of baseline data, and the monitoring to be performed during and after any activities in the exploration area with potential to cause serious harm to the environment. They were based on the recommendations of an international workshop convened by the ISA in 1998 which had recognized the need for clear and common methods of environmental characterization based on scientific principles and taking into account oceanographic constraints.

The 2001 recommendations were revised in 2010 in the light of increased understanding. Then, in 2013, following the adoption of exploration regulations for sulphides and crusts, the Commission decided that there was a need to create a comprehensive set of environmental assessment guidelines that dealt with all three types of marine minerals. The revised and updated recommendations take into account new knowledge, including the outcomes of workshops convened by ISA, and set out the detailed observations and measurements that need to be made while performing specific activities and recommended data collection, reporting and archiving protocols. The recommendations are accompanied by a glossary of key terms and an explanatory commentary.

Importantly, the recommendations also elaborate on and clarify the obligation on contractors to undertake environmental impact assessments (EIA) by listing the activities that do and do not require prior EIA. The threshold where EIA is required is generally around the point at which the contractor begins at-sea testing of collecting and processing systems. With respect to those activities that do require a prior EIA, the recommendations require the contractor to submit the EIA and a proposed monitoring programme to the Authority at least one year before the activity takes place. The recommendations include a reporting template for the EIA, as well as details of the observations and measurements to be made during and after the activity in question.

For longer term planning, the Authority played an important role in developing the first ever regional environment management plan for the Clarion Clipperton zone. This is regarded as so important that the General Assembly of the UN has called upon the Authority to develop similar environmental management plans in other regions where significant exploration activity is taking place.

Environmental regulation during the exploitation phase
It is obvious that environmental regulation will need to be far more stringent during the exploitation phase. This is because certain effects from deep sea mining are already known to be harmful to the marine environment as tentatively listed during the first EcoDeep-SIP workshop.

Therefore, the Authority is now in the process of developing exploitation regulations, which is a challenging and complex exercise. The process began in 2013 with the issue of a technical Study on the issues associated with the development of a regulatory regime for mining. Subsequently, the legal and Technical Commission conducted a public consultation in which it sought the views of stakeholders as to some of the key policy issues arising in connection with exploitation. In July 2016, the same Commission issued a 'zero draft' of the exploitation regulations. This was made available for public consultation until November 2016. The results of that consultation were considered by the Commission at its recent meetings in Kingston and it is expected that a further report will be issued to the Council in August 2017.

The workshop reports and working documents issued by the Authority raise numerous issues that will need to be considered as the process moves forward:

- What should be the content of the EIA process?
- How prescriptive should the regulations be? Should they set minimum standards or should they be comprehensive? Should there be a reference standard, e.g. Good Industry Practice? How is this defined?
- What is the EIA measured against? What constitutes an acceptable baseline from a legal and scientific perspective? Who determines whether the baseline data are sufficient (Legal and Technical Commission, contractor, external peer review?)
- What are the requirements for an EIA report?
- What is the content of an Environmental Management and Monitoring Plan? Should it cover the entire contract area or can it be delivered according to a progressive mining plan?
- What is the process for review and evaluation of EIA and Environmental Management and Monitoring Plans?
- How will environmental targets and thresholds be determined for the purposes of environmental monitoring?
- To what extent should information be publicly available?
- To what extent should there be public participation in decision-making?
- How does project-specific environmental management relate to regional environmental management and vice-versa?

Concluding remarks

From the above, the role being played and to be played by the scientific community in regulatory development and on-going environmental decision-making is crucial. The Authority seeks to work more closely with the scientific community to help define priorities for marine scientific research to improve environment baseline studies. We also look to the scientific community to help develop adequate environmental monitoring and management programmes to assess the future effect of mining activities on the marine environment.

In light of the current phase of development of deep seabed mining, we are presented with a unique opportunity for the scientific community to work in partnership with the Authority towards a new phase of ocean science and exploration to meet the challenges of the twenty-first century.

Regional Reports

Deep seabed minerals development in the Pacific Islands Region – Current status and challenges

Akuila Tawake, Head of Geo-surveys and Geo-resources Sector, SPC

The Deep Sea Minerals Project is a collaboration between the Pacific Community (SPC) and the European Union (EU). Initiated in 2011, the €4.4 million DSM Project is helping Pacific Island countries to improve the governance and management of their deep-sea minerals resources in accordance with international law, with particular attention to the protection of the marine environment and securing equitable financial arrangements for Pacific Island countries and their people.

The DSM Project has 15 member Pacific Island Countries: the Cook Islands, Federated States of Micronesia, Fiji, Kiribati, Marshall Islands, Nauru, Niue, Palau, Papua New Guinea, Samoa, Solomon Islands, Timor Leste, Tonga, Tuvalu and Vanuatu.

A primary objective of the project is to support informed and careful governance of any deep sea mining activities in accordance with international law, with particular attention to the protection of the marine environment and securing equitable financial arrangements for Pacific Island countries and their people. The Project is also working to encourage and support participatory decision-making in the governance and management of national deep sea minerals resources. Despite the rapidly growing commercial interest in deep sea minerals many Pacific nations do not have the necessary legal or management systems needed to ensure the responsible management of these important natural resources.

In response to this urgent need Pacific Island countries requested the development of a regional project to help governments develop the national frameworks and technical capacity needed to strengthen the management of their national deep sea mineral resources.

As a result, since its inception (2011), the project carried out many studies and case studies including environmental, financial, scientific research management frameworks, costs and benefits assessment, environmental management needs, and a regional legislative and regulatory framework now promoted by the South Pacific Commission towards all its countries and territories members.

More particularly, the Regional Financial Framework for Deep Sea Minerals Exploration and Exploitation is aimed at providing Pacific countries with a guide to the major issues to be addressed when setting up national financial frameworks. It was prepared in collaboration with the International Monetary Fund through the Pacific Financial Technical Assistance Centre and provides an overview of key issues in the financial management of revenues and wealth associated with the potential development of deep sea minerals in the region.

The Regional Environmental Management Framework for Deep Sea Minerals Exploration and Exploitation contains an overview of deep sea mineral deposit environments and potential environmental impacts of deep sea mining projects, as well as management and mitigation strategies, including an environmental impact assessment report template.

The environmental framework serves as a guide for Pacific countries, informing and supporting them to make sound decisions regarding their deep sea mineral resources and to take appropriate measures to reduce environmental risks, should they wish to engage in the mining industry.

Anticipating, assessing and governing the social and environmental impact of deep-sea mining – Prospects from French Polynesia

Pierre-Yves Le Meur, Anthropologist, IRD-GRED, Montpellier, France

Following a joint request from the French State and the French Polynesia (FP) Government, the Research Institute for Development (IRD) was commissioned to coordinate a 'collective expertise' on the prospect of deep-sea mineral resources in FP's huge exclusive economic zone (EEZ). The main reasons underpinning this request were about investigating what could be a new resource and economic activity in the future in a region where development opportunities are scarce, and a public will to understand and act accordingly.

The IRD expert group review was a robust and interdisciplinary methodological model capitalizing on 14 expert group reviews. It consisted of assessing the state of existing knowledge (natural and social sciences) in the region, pondering the different points of view (governments, market, civil society), and coming up with synthetic conclusion and recommendations for the future.

This study was thus unique in the sense it allowed to develop an early strategy and policy, identify and partly filling knowledge gaps (resource, technology, environment), paving the way to a democratic and inclusive governance (FPCI).

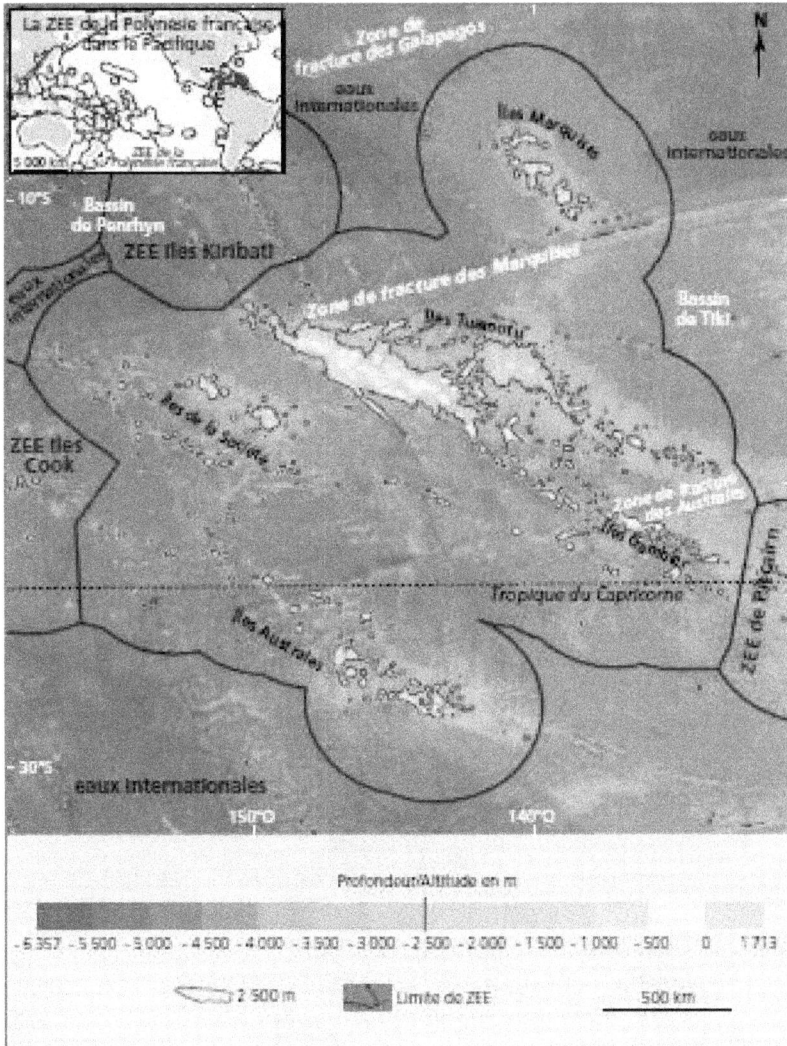

Figure 1: French Polynesia EEZ

The questions the study tried to answer were about the state of knowledge from a general point of view (protocols, methods, available technologies, and associated impacts), mid- and long-term issues at stake (mapping, pilot sites, investments, commodity chain, local benefits, risks), and the policy options leading to a possible master plan for sustainable exploitation.

The main conclusions were as follows:

- World-class potential of the cobalt-rich crusts on seamounts and phosphate substratum;
- They are many uncertainties and specific risks involved in their exploitation;
- There is therefore an urgent need to deepen knowledge of these resources and associated living communities;
- Sustained by an explicit and tailored deep-sea mineral resources policy.

The main policy issues in valuing this potential are to be considered now, before any exploration/exploitation take place. An early and adapted policy is therefore needed for managing heterogeneous temporalities (geological, political, ecological, cultural) as well as associated risks and uncertainties from technological, environmental, economic, normative, and ethic points of view, with the following set of recommendations:

STOP OR GO?

R1.   Create an information system to impose consistency and organize access to existing data.

R2.   Instigate programmes to generate knowledge and develop suitable technology.

R3.   Define a development strategy for a deep-sea mining industry or, alternatively, decide to abandon the project, after combining acquired data with detailed technical and economic scenarios as well as initial consultations.

IN THE CASE OF 'GO'

R4. Organize governance at a sufficiently early stage and with the participation of all stakeholders.

R5. Involve the country in regional, European and international initiatives to further the development of marine mineral resources, including cooperation, research and innovation with partners like Japan.

R6. Conduct research and development programmes focused on innovative technology in the fields of exploration, mining and metallurgy of deep-sea mineral resources.

R7. Create efficient and attractive administrative and regulatory instruments to develop a deep-sea mining sector.

R8. Set standards for selecting, monitoring and evaluating mining projects in order to promote control and transparency in public communication.

R9. Organize the monitoring and evaluation of the deep-sea mineral resources policy in order to measure its effects and, when necessary, make changes.

As regards the impact assessment processes, social impact is most of the time the 'orphan of the assessment process' though it will be crucial for any deep-sea mining initiative (exploration and exploitation) to take into account peoples' knowledge, representations, values, and uses of maritime areas.

To do this, the different time dimensions must be taken into consideration in between geological, cultural, economic, and political temporalities. It is about setting up a sustainable management strategy for the deep ocean establishing science-based conservation goals, developing an overall framework for defining baseline conditions, and establishing monitoring requirements for follow up and accountability.

Managing impacts of deep sea resources exploitation – Lessons from the EU-funded project MIDAS

Jean-Marc Daniel, IFREMER, Director, Department of Physical Resources and Deep-Sea Ecosystems[1]

MIDAS (Managing Impacts of Deep-Sea Resource Exploitation) was a 3-year EU-funded project (2013-2016) gathering 32 European research organisations (including IFREMER) with a total budget of 12 Million Euros. The project focused mainly on the potential impacts associated with extraction of manganese nodules and seafloor massive sulfides (SMS), but also addressed environmental issues related to the exploitation of methane gas hydrates, and the potential of deep-sea muds in the North Atlantic as a source of rare earth elements (REEs).

Its main objectives were:

1. *Identification of the scale of possible impacts, and their duration, on deep-sea ecosystems associated with different types or resource extraction activities;*
2. *Development of workable solutions and best practice codes for environmentally responsible and socially acceptable commercial activities;*
3. *Development of robust and cost-effective techniques for monitoring the impacts of mineral exploitation and the subsequent recovery of ecosystems;*
4. *Work with policy makers to enshrine best practice in international and national regulations and overarching legal frameworks.*

MIDAS project's timing was very opportune since, as previously presented, it coincided with the ISA's development of a mining code for exploitation of deep-sea minerals.

---

[1] More detailed facts can be downloaded from MIDAS two publications: 'Managing impacts of deep sea resource exploitation – Research highlights', and 'Implications of MIDAS results for policy makers: recommendations for future regulations'. www.eu-midas.net

MIDAS 9 studied oceanic sites included the mid-Atlantic Ridge (SMS), the Clarion Clipperton Zone (CCZ) of the Central Pacific (nodules), the Black Sea, the Norwegian and Svalbard continental margins (gas hydrates), whilst the Canary Islands, palinuro Seamount (central Mediterranean), Norwegian fjords and portman Bay in Spain were used as proxy sites for various mining impact experiments. In addition to 30 research expeditions to these areas, a collaboration with the European Joint Programming Initiative (JPI) Oceans pilot action "Ecological aspects of deep-sea mining" enabled the project to work jointly on data from three expeditions to the CCZ and Peru basin.

MIDAS Work Programme

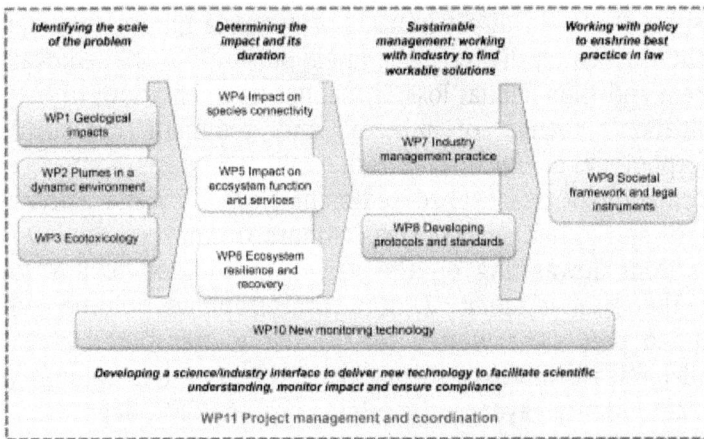

As shown above, the scientific work was divided into the examination of the scale of the potential impacts, and how these impacts would affect ecosystems regarding connectivity between populations, ecosystem functioning, resilience and ability to recover once mining ceased. Since partners were also from the industry, information about likely mining scenarios for adapted protocols and standards could be gathered from the private sector. The combination of new scientific data with projected mining scenarios and accepted best practice enabled the project to put forward an environmental management framework facilitating responsible mining whilst taking into account environmental concerns. A social dimension was incorporated into the approach through close engagement with civil society. Finally, monitoring technology was identified, indicating which technology is currently available and which requires further development.

MIDAS scientific work addressed the scale of the potential impacts from deep-sea mining, to begin with the size of the areas to be mined. Interestingly, while the areas required per million tons mined compared to land-based mining are much larger in the case of polymetallic nodules and cobalt crust, it is the opposite when considering massive sulfides (SMS) mining (0.054 km^2 versus 0.12 km^2 on land).

To complement the observations made at the first EcoDeep-SIP workshop, the main direct impacts of deep-sea mining include: 1) mortality of fauna living on mined substrates; 2) removal of substrate and thus habitat loss; 3) habitat fragmentation; 4) habitat modification (change of mineral and sediment composition, geomorphology, chemical regimes). Indirect impacts comprise: 5) the formation of near-seabed sediment plumes by the activity of crawlers and other seabed installations; 6) the returned water plume from dewatering on the vessel, in addition to any leaks along the riser system; 7) the trans-shipment plume when dewatered ores are rewetted for transfer to transport barges; 8) potential release of toxic substances into the water column and Benthic Boundary Layer by the mining process (SMS deposits).

MIDAS findings indicate that the loss of specific habitats within some areas (nodule fields and inactive hydrothermal vents) will persist in the long term; for nodule fields this change can be considered permanent for nodule-attached fauna owing to the very slow rate of nodule growth. The impacts of direct habitat loss are compounded by long-term changes to physical conditions, such as altered sediment structure, caused by mining and re-sedimentation.

There is variation in recovery rates after disturbance for different species of fauna found within nodule- and vent-associated habitats: some species are found to colonise sites within a few years while others have not been recorded returning to sites after more than 26 years (French claim in the Clarion Clipperton Zone). It should also be noted that observed recovery rates are mostly known from relatively small-scale disturbance experiments; the potential for deep-sea environments to recover after disturbance on the scale of full-scale mining operations is not known.

Communities of sessile fauna in nodule-rich areas are similar across the Clarion Clipperton Zone (CCZ), but the nodule-poor areas show much lower diversity and abundance of fauna typically found on nodules. Work by the EU Joint Programme Initiative, Oceans (JPIO) project and related MIDAS research highlighted the importance of nodules in maintaining epifaunal biodiversity in the CCZ. It is therefore expected that recovery will be faster and greater in the higher productivity areas of the eastern and central CCZ relative to the lower productivity western areas. Communities of mobile fauna show the same trends. These finding have consequences for the designation of potential preservation reference zones in the CCZ and should be incorporated into conservation management plans.

Beyond substrate removal, particle plumes may have significantly adverse impacts: although they will not be apparent at the sea surface, capped by density stratification, they will be difficult to meaningfully map in three dimensions in the deep-sea environment. Therefore, accurate models are vital tools for predicting and understanding plume impact. In the case of MIDAS' experiments, plume modelling has taken a particle-tracking approach in which near-seabed plumes generated by the mining process and mid-water plumes generated by dewatering of the ores at the sea surface are represented as a large number of individual particles of different sizes and settling velocities.

In the case of hydrothermal site and massive sulfides on the Mid-Atlantic Ridge, models predict that particles will disperse over a large depth range, more particularly for the finest particles, due to the fact that the deep sea may be highly turbulent because of density stratification and near-bed current speeds remotely driven by the passage of eddies sometimes generated thousands of kilometers away.

In the case of abyssal nodule fields, particles are predict to disperse well beyond the area that is mined with a pattern of plume deposition highly directional following the near-bed currents.

Plumes within the water column may form sinuous patterns as they are stretched and stirred by eddies and other flow structures. For this reason, it is expected that the monitoring of plumes within the water column will reveal great patchiness in the shape and extent of the plume. Again, it is why models are key to designing and interpreting monitoring of full-scale mining operations.

As regards the ecotoxicology, MIDAS results show that it will not be possible to extrapolate toxicity information and practices from shallow water areas to deep-sea situations. Under controlled ecologically-relevant conditions it may be possible, using appropriate bio-essays, to determine the bulk lethal toxicity of an ore deposit and of the return waters using a number of different biological proxy organisms. Consequently, the bulk resource should be processed as anticipated during exploitation, and the toxicity of the resulting aqueous solution and particulate material should be assessed both independently and in combination under ecologically-relevant environmental conditions (temperature, hydrostatic pressure, oxygen concentration, carbon dioxide concentration over time-scale representing a range of acute to chronic seasonal and reproductive cycles) timescales for key species present in the area of impact.

In regard to the benthic and returned water plumes, one of the issues is about who will be responsible for measurement in the license block, in adjacent license blocks and unlicensed areas in order to quantify the impacts and their spatial extent and ensure these lie within the acceptable levels agreed between contractors and regulating bodies.

The selection of proper monitoring tools and strategies need not only to consider these temporal and spatial scales of potential impacts but also to take into account the characteristics of the local environmental and ecological conditions, as well as the expected timescales of recovery. Existing or new technologies are therefore needed in routine ecosystem monitoring in contract areas as well as Areas of Particular Environmental Interest (APEIs).

One of the new technologies is about 'Integrative habitat mapping for identification of different deep-sea habitats and their spatial coverage'. AUV-based seafloor surveys with high resolution side-scan sonar and a novel fast camera system were successfully carried out in nodule ecosystems of the tropical eastern Pacific. Progress has been made as well for the development and use of software tools for the analysis of seafloor imagery collected from potential mining sites. In the future, non-invasive habitat mapping technologies offer great potential for the further development of automated systems to monitor the environment in the industrial setting.

Regarding biogeochemical monitoring, as an example, a suite of autonomous landers and ROV modules equipped with enclosures and micro-sensors was deployed in nodule ecosystems of Peru Basin and provided strong indications of the long-term impacts of simulated mining disturbances on biogeochemical processes at the seafloor.

The MIDAS report 'Tools for rapid biodiversity monitoring across size classes' focuses on the potential of novel image-based and molecular technologies to speed up biodiversity assessments and to be used for routine application by industry. Before that, a detailed database of voucher specimens from each concerned region, their morphology and genomic sequences, and their appearance in images will need to be set up.

The identification of technologies that are best suited for use in the context of routine industrial monitoring was a key MIDAS goal through compilation and assessment of available tools:

1) description of available technologies (imaging surveys with different platforms, molecular assessment of micro-, meio-, macrofauna including environmental/extracellular DNA);
2) strength-weakness-opportunities-threats (SWOT) analysis and evaluation of methods with common criteria (size-classes addressed, taxonomic resolution, spatio-temporal coverage, readiness, costs, etc.); and
3) case studies from MIDAS and associated projects.

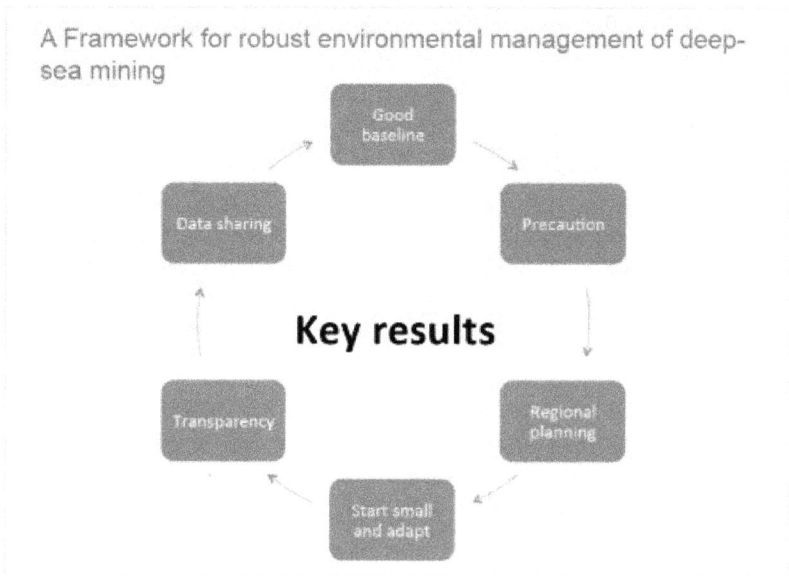

A Framework for robust environmental management of deep-sea mining

Good baseline · Precaution · Regional planning · Start small and adapt · Transparency · Data sharing

Key results

A robust environmental management framework for deep-sea mining has been proposed. It is mainly based on the precautionary approach incorporating adaptive management into its design and focusing on the phases of a single mining project where regional and claim-scale management issues are integrated. In MIDAS' view, 'the adoption of such an environmental management framework by the ISA and national regulators for deep-sea mining would have three main benefits: i) managing impacts from an individual project as well as cumulative impacts of multiple projects; ii) reducing uncertainty in planning, applications, and undertaking exploitation and extraction activities; iii) fairness and uniformity in the application of environmental standards, with equal responsibility and liability between contractors.

In all, the three years of scientific and technological studies by MIDAS have led to significant knowledge and understanding to support the development of environmentally and socially responsible seabed mining regulations. The results have confirmed the importance of broad-scale regional environmental management planning, as well as the need for more finely tuned site-specific management of mining areas consistent with the broader plan.

MIDAS has also generated new on-going projects like 'Bluenodules' (http://www.blue-nodules.eu/), 'Blue mining' (http://www.bluemining.eu/), or through the Joint Programming Initiative Oceans (JPI-Oceans).

Session 1

DAY1: 14 MARCH 2017

Chair: Chair:Yves Henocque (IFREMER)

Vice-Chair: Tomohiko Fukushima, JAMSTEC and
Makoto Seta (Yokohama City University)

1.1 Objectivity and reliability of data

Brief Orientation
Tomohiko Fukushima (JAMSTEC)

In order to conduct proper environmental impact assessment (EIA), rigorous and reliable data-sets are indispensable. In the case of ocean physics and chemistry, provided research instruments fast development, accuracy and precision are becoming higher than in the past. Instead, in the field of biological identification, reliability of data-sets is depending on a few taxonomic experts. However, when deep sea mineral resources development will really take off, it will become impossible for the few experts involved to deal with the resulting very large number of samples. In such a case, it is not unlikely that there will be a lack of expert to perform biological identification. In addition, because a data cross-check system is not established so far, the spreading of incorrect biological data can be of concern. In such a context, in order to prevent the spreading of those incorrect data, the International Seabed Authority (ISA) has already convened three taxonomic workshops with great appreciation from the participants. Taking this situation into account, Japan SIP is working on providing new technologies for producing accurate and reliable data to perform trustable EIA.

Introduction of Metagenomics
Miyuki Nishijima (JAMSTEC)

For environmental impact assessments of plans for seabed mineral resource exploitation, objective, comprehensive and easy-to-apply analysis techniques are required. Traditionally, relatively large organisms have been used as indicators for environmental impact assessments, and analysis based on morphological observation has been conducted. Also, not enough research has yet been carried out on the organisms that inhabit the deep seabed. On the other hand, the development of analytical methods by molecular biology in recent years is remarkable, and the introduction of the next generation sequencer has made it possible to analyze a large number of gene sequences. Meiofauna, being ubiquitous as well as sensitive to environmental perturbations, have been chosen as indicator organisms for our metagenomic analyses. In this presentation, we will introduce 18S rRNA amplicon sequence analysis for meiofauna by next generation sequencer.

Discussion feedback

User Opinion:
Analysis seems powerful and time-saving but technology hurdle seems great for South Pacific Island states. Need help with capacity building. Regional approach better? Very little (none?) DNA barcoding-taxonomy efforts in region so a GenBank search would probably get no matches.

Administrator Opinion:
How can this very specific taxonomic analysis inform us about the ecosystem-environment, which is what we the users want to know about? Can meiofauna species presence data be informative enough? Even singletons need to be barcoded and have taxonomic work done on them, though they may not be included in the final community analysis. Sequences without taxonomy are useless.

Other points:
Japanese taxonomists could teach University courses for capacity building – funded by Japanese government? JAMSTEC specialists to collaborate with South Pacific University and Papua New Guinea.

An ISO standard protocol for meiofauna community analysis using metagenomics is being worked on. Database metadata formats need to be standardized also.

1.2 Working efficiency

Brief Orientation
Tomohiko Fukushima (JAMSTEC)

Deep sea mineral resources development is heading towards a probable new economic activities. So that working efficiency is required first of all regarding EIAs comprehensive and accurate procedure. However, meiobenthos, which is one of the main components in benthic community, is rather small ($>32\mu m$, $<250\mu m$) and, when sampled, is always sticking to fine particles of sediment. Therefore, the working process of sampling and identification must be carried out carefully, which is incompatible with speedy efficiency. Moreover, due to organisms' small size, additional care is required such as sectioning samples, dyeing, sieving, and observing them under the microscope. Besides, the International Seabed Authority (ISA) recommended to contractors that the number of samples must be large enough to understand abundance and biomass for robust statistical analysis. A workshop convened by ISA has requested that enough samples should be taken to understand species diversity in their survey site. As mentioned above, considering they are delicate samples, requiring a complicated procedure, with a large number of samples to be treated, working efficiency will be an indispensable task for commercial mining. On such background, Japan SIP will propose a technique that includes automatic sorting and photographic systems.

A rapid method to analyze meiofaunal assemblages using an Imaging Flow Cytometer
Tomo Kitahashi (JAMSTEC)

Meiofauna are usually defined as benthic organisms that pass through a 500–1,000 µm sieve and are retained on a 32–63 µm sieve. Compared to larger macrofauna and megafauna, meiofauna in the deep sea have high abundance and biomass and are an important component of deep-sea ecosystems (Wei et al., 2010). In addition, meiofauna have considerable influence on the nutrient cycling in the sediments and sediment stability. Meiofauna are widely recognized as a useful indicator for assessing the effect of anthropogenic and natural disturbances on the deep-sea ecosystem (Giere, 2009; Zeppilli et al., 2015). However, traditional methods of investigating meiofauna, which include individually counting and identifying small-sized meiofaunal specimens under a microscope, are labor-intensive and time-consuming. In addition, advanced expertise is required for the identification of meiofauna to the species, genus, or even family level, and the number of qualified to do this is limited. Also, if a technician does not have the training or knowledge to identify meiofauna, the dissemination of inaccurate data could result. Alternative methods, which can rapidly process a high volume of meiofaunal samples, are required when conducting long-term environmental impact assessments (EIA, including rapid detection of drastic change and subsequent monitoring of environment). Taking this situation into account, SIP also provide a novel technique for analyzing meiofauna assemblages for proper EIA.

Discussion feedback

User Opinion:
Technology hurdle seems too great for South Pacific Island states but a Regional approach with University of the South Pacific and the Pacific Island University Research Network might work. Japanese taxonomists to teach course either at USP or in English at a Japanese University? Include statistical-community analysis methodology in the Protocol would be helpful.

Administrator Opinion:
Would be good to groundtruth the method where the community is well known so we could be sure of its worth.

Other points:
Citizen science could help raise the accuracy of the Automatic Image Recognition process. Need an online database/server for image upload and AIR analysis. Who would provide this

infrastructure and maintain it?

Possibility to make image+DNA sequence database if effect of rose bengal dye on PCR amplification can be countered.

1.3   Technical Feasibility

Brief Orientation
Tomohiko Fukushima (JAMSTEC)

The Area and its resources are 'common heritage of mankind'. Therefore, any technical feasibility should rely on rules and guidelines in regard to deep-sea mineral resources mining. Like for any other activity, EIA is indispensable and compulsory. However, current seafloor observation technologies cannot match the level of recommendations issued by ISA. In general, annual fluxes onto the sea bottom are not constant, and some of the deep sea benthic organisms synchronize their reproduction rhythms to those flux fluctuations. The fact that benthic organisms' abundance has also seasonal variations, suggests that a once a year baseline survey is inadequate. The International Seabed Authority (ISA) recommends taking into consideration the environment natural variability. More particularly in regard to the seafloor, continuous camera observation system is recommended at least over one year. However, for the time being, equipment that complies with this request is hardly available. Besides, multiple cruises over years will put an excessive burden on contractors. Taking this situation into consideration, Japan SIP is developing a long-term submarine observation system that can meet ISA recommendations. This system is devised to be lightweight, of compact size, and made of ready-made parts so that it can be used both by pioneer investors who have been working in the field for a long time, and new entrants.

Introduction of the Edokko Mark I
Tetsuya Miwa (JAMSTEC)

The secretariat asked the lecturer for technical commentary on "Long-Term Monitoring Images near the Deep Seafloor by Edokko Mark I".

In the lecture, explanations such as compact body, lightweight and cuatomizable etc. about this system are expected.

Discussion feedback

User Opinion
Cost is not so high, making the instrument so attractive! Running cost low also because easy maintenance and can be deployed from small ships. Would be good to also provide tools to help management/analysis/storage of data as a set/system

Administrator Opinion
Would like long term data from sensors like CTD etc. so protocol to get maximum return for battery life would be useful.

Other points
Interested to see results for deployment of multiple Edokkos at the same time to assess whether useful for estimating variability and to increase number of megafauna observations to allow statistical studies.

Possible to custom design larger system to get maximum data return for a given ship size or other operational constraints.

Session 2

DAY2: 15 MARCH 2017

Chair: Jean-Marc Daniel (IFREMER)

Vice-Chair: Masanobu Kawachi (NIES)

Rapporteur: Dhugal J. Lindsay (JAMSTEC)
and Makoto Seta (Yokohama City University)

2.1 Unexpected impact

Environmental pollution monitoring system using bioassay (brief orientation and technical explanation)
Masanobu Kawachi, Hiroshi Koshikawa (NIES)

Seafloor mining operations, including transport of the mineral ores from the seafloor to mining vessel, would be well-designed to minimize environmental impacts. During the mining operations, there remain some risks of leakage of the ores and the subsidiary metal-contaminated seawaters to the marine ecosystem. Although potential impacts for surface discharge were concerned in the ISBA/19/LTC/8 and other guidelines, any effective methods to assess the harmful environmental effects are still unclear and would be urgently developed.

We think "bioassay" is one of the effective technologies for that purpose. The water quality management systems with "bioassay" such as WET, DTA, WEA are recently becoming popular worldwide especially for evaluating and regulating effluents from factories/sewage plants. Besides, the latest MIDAS report (2016) for policy makers and future regulations raise a subject on ecotoxicology among the EIA issue.

In this session, firstly, we introduce our scientific research results about the release potential of metals and metalloids from mineral particulates and the impact of the leaching metals on marine phytoplankton communities. Secondly, we propose a small and rapid bioassay system, though it is still under development, as the onboard seawater safety monitoring system which can detect the unexpected pollution at seafloor mining sites.

Discussion feedback

User Opinion
Papua New Guinea should do this kind of baseline study because they are further along development path than other South Pacific

States. Both PNG and Suva have the lab equipment already. PNG (not yet other SP states) should work on monitoring too. Who does monitoring? Who does reporting? Definitions need to be addressed.

Administrator Opinion

Not clear why Manganese was not detected during leaching experiment. Some problems with methodology? Use of artificial seawater may be problematic because complex-forming materials absent? Very little manganese present in the ore that was used for the leaching experiment. Land mining toxicity studies target humans but need new standards for marine systems.

Other points

Cyanodium sp. lives in oligotrophic areas where mining is proposed and can be stored frozen and brought back to life.

Toxicity in high oxygen environments greatest. Delayed Fluorescence test is good proxy.

Coolant water intake on ships cheapest way to get water samples but may be contaminated from ship pipes. Low level contamination hardest to detect but want to try. Frequency of sampling should be once per day but also need an accident detection system, which we are working on also. Can be applied for land-based spills also.

2.2 Potential technology

Brief Orientation

Hiroyuki Yamamoto (JAMSTEC)

The physical oceanography is an indispensable factor for assessment of marine ecosystem. The ocean current, tide, down- or up-welling, and turbulent flow in marine environments strongly affect spatial condition of habitat, migration of organisms, and dispersion of planktonic larva. The water flow conducts the migration opportunity and dispersion rate, and sustains the robustness of biodiversity. The impact assessment on seabed mining needs the information on water flow on local and regional scale. Possibility of biodiversity recovery after the exploitation impact can be estimated by the potential of migration and dispersion among communities of region scale, e.g. Okinawa Trough. Effect of a plume caused by mining activity which extends to horizontal and vertical directions, can be estimated by the water flow and physical structure in local scale, e.g. knoll and seamount. Observation techniques and tools for local scale determination are different from reginal and global scale. We should consider such technical issue and make suitable research and monitoring plan for assessment and management.

Turbulence measurement for plume diffusion estimation
Yasuo Furushima (JAMSTEC)

Seafloor disturbance and subsequent suspension, diffusion and re-deposition of sediment particles caused by submarine mining negatively affect habitat and ecosystem of benthos. Understanding fluid mechanism which predominant the particle behaviors near deep-sea bottom is important to assess environmental impacts of submarine mining. However, bottom boundary turbulence is remarkably complicated due to hydrothermal vents and surrounding unique bottom topography.

Direct measurement of near-bottom turbulent flow is essential to clarify the fluid dynamics in deep sea. However, there are extremely few available observations because the direct measurement technique has been established very recently. This presentation introduces the direct measurement of the deep sea turbulent flow as one of potential technologies for environmental impacts assessment of submarine mining, and its application study on oceanic turbulent flow and suspended particle behavior in deep-sea bottom boundary layer.

Discussion feedback

User Opinion

Would like any model to incorporate possible particle distributions from effluent discharged at surface or at other mid-depths also. Would like more information on the effect of the local bathymetry on near-bottom mixed layer thickness. SPC has ADCP and multibeam survey equipment as well as staff who could do this kind of study although have never used this kind of turbulence meter before. For deep-sea mining the private sector would do this kind of study with SPC focused on shallower areas or fished areas. Tie-ups with Australian and New Zealand universities under consideration also for monitoring.

Administrator Opinion

Current direction information is also important for determining where the SMS deposits should be most abundant. Most importantly, there is a need for long-term series of data through long-term observation system.

Other points

The VMP-X turbidity probe does not incorporate a conductivity/salinity sensor so density measurements must be made by concurrent X-CTD probes or CTD casts.

Trying to determine useful proxies for assessing particle distributions where measurements are easier. Also need development of a sensor to measure turbidity very close to bottom without churning up sediment.

Session 3

Chair:Yves Henocque and Jean-Marc Daniel, IFREMER
Vice Chair: Tomohiko Fukushima, JAMSTEC and
Masanobu Kawachi, NIES

Closing

The two co-chairs conveyed their gratitude to the organizers and participants for the workshop impeccable preparation and organization and for the quality of the presentations and related discussion. They emphasized that the dialogue thus generated between institutions (private, national and international) is key, more particularly because we have the chance to embark early into the process, time to anticipate, before actual exploitation.

Deep sea mineral research and development should not be considered in isolation but as one of the maritime activities (current and potential) under the common goals set by the Sustainable Development Goals (SDGs) and more particularly SDG14. This invites us into different ways of thinking broadening the scope to interdisciplinary reflection (natural and social sciences) and, generally speaking, participation of local governments and populations (including local knowledge) into the decision-making process.

It is also reminded that this workshop is the second SIP-EcoDeep workshop (the previous was held in June 2015) and, as underlined by an ISA representative, already big improvements have been putting in adequacy the ISA requirements and technological development.

Regarding the latter (technological development), it is also reminded that the underpinning rationale is to develop technologies and methods that might be easily made available to developed AND developing countries hence the importance of sustaining a dialogue between administrators (ISA, regional and national administrations), research institutions (e.g., JAMSTEC, IFREMER), and the private sector (providers and users) like this workshop has intended to do, more particularly in regard to the Asia Pacific region.

The environmental impact assessment and monitoring process is thus a complex one that requires many different disciplines and kind of technological development that will need to be further integrated and running on the long term (long-term observation) to fit in the ISA elaborated requirements (IBSAs). That could be the very subject of a 3rd SIP-EcoDeep workshop with the help of its supporting organizations.

Recommendations

The main recommendations may be summarized as follow:

Promote and transfer technological development

- Clearly indicate technologies state of development: which are mature, ready to use, and which are under development.
- A complete set of technologies, as some have been presented in this workshop, should be applied and tested at a single site to see how they can be used together.
- An additional recommendation has been made towards ISA to also contribute to the Technology Development efforts though appropriate fund sources have still to be identified.

Promote long-term observation system for getting long-time series of data

- All the scientific work and techniques introduced during the workshop were recognized as totally appropriate in regard to ISA guidelines and directives, though measurement through long-time series is now needed like, for example, sediment oxygen profiles over time as a proxy for biological activity.

Be strategic and communicate knowledge

- Common goals following the SDGs are not being set between policy makers, scientists and business at present. They need to be built up on a common understanding of mining benefits and environmental losses.
- Communication of knowledge as well as uncertainty and dialogue with local governments and people are key. The role of 'science communicators' is crucial and should be promoted as well as interdisciplinary reflection including social sciences.
- Policy-science interface needs to be rethought? Is citizen science a good complementary tool and at what conditions? The same definitions hence understanding need to be used by scientists and policy makers.
- Scientific data/results are the vocabulary for communicating if something is "safe" or not and private companies, with the help of comprehensive Environmental Impact Studies (EIA), could make/use of this kind of data and information to dialogue and convince locals about possible mining.
- Local communities can be an important source of information for scientists engaged in EIA processes and studies since they often have deep knowledge of the marine environment.

Promote capacity building at regional scale through national institutions

- The ISA Secretary General confirmed how much he was impressed by Japan's commitment to technology development and rejoiced to see that ISA recommendations were tightly followed. Certainly, one of the key conditions for future adoption and application will be thorough and continuous capacity-building that could be helped through ISA Endowment fund for Japan and France whilst scientific institutions like JAMSTEC and IFREMER will take in charge the actual training.

- Fiji and PNG can collaborate with foreign universities and research institutes in their own right but most other island states don't have the resources. Trained workers tend to migrate to Australia, New Zealand and the USA so always need to keep working on capacity building. Future **capacity building** should concentrate on Regional rather than National scales. For that purpose, the University of the South Pacific is a good partner.

<u>Sustain dialogue towards practical improvements</u>: to advance further in implementing all previous remarks and recommendations, face-to-face informal working meetings such as this workshop are considered crucial in the future.

Guests and Speakers

SIP Program Director and Professor Emeritus of the University of Tokyo

Tetsuro URABE

Dr. Tetsuro Urabe, born in 1949 in Okayama, Japan, graduated from the University of Tokyo (UT) in 1971 and received a Ph.D. in geology at the UT in 1976. His professional experience includes Assistant Professor, Geological Institute, UT (1976 - 1985), Visiting Researcher, Department of Geology, University of Toronto, Canada (1979-1981), Senior Research Geologist and the Chief Geologist, Geological Survey of Japan (1985 - 2000), before he became Professor, Department of Earth and Planetary Science, UT (2000 - 2014). He is currently Professor Emeritus of UT, Visiting Professor of Kyushu University, and executive adviser at the Mining Engineering & Training Center. He currently serves as the Program Director of SIP "Zipangu in the Ocean Program", a member of Commission on the Limit of the Continental Shelf of United Nations Convention on the Law of the Sea (2011-2017), and Councilor for The Ministry of Foreign Affairs of Japan. He led several major sea-going research projects as PI, which include; (1) Japan-US "Ridge Flux Project" (1993-1998) on super-fast-spreading East Pacific Rise; (2) "Archean Park Project" (2000-2006) on deep biosphere related to seafloor hydrothermal activity, and (3)"TAIGA Project" on Trans-crustal Advection and In-situ biogeochemical processes of Global sub-seafloor Aquifer (2008-2012).

Executive Director, JAMSTEC

Yoshihisa SHIRAYAMA

Dr. Yoshihisa Shirayama, born in 1955 in Tokyo, Japan, obtained D. Sc. Degree from Graduate School of Science, The University of Tokyo (UT), in 1982. He then served Assistant and then Associate Professor at Ocean Research Institute, UT. In 1997, he became a professor of Seto Marine Biological Laboratory, Faculty of Science, Kyoto University. In 2003, the laboratory moved to Field Science Education and Research Center. He served as Director of the center from 2007. In April 2011, he became Executive Director of Research, Japan Agency for Marine-Earth Science and Technology. His major research field is marine biology, especially taxonomy and ecology of deep-sea meiobenthos. He also is working on the marine biodiversity and the impact of ocean acidification upon it. He was awarded "Okada Prize" from Oceanographic Society of Japan in 1988, Minister of Environment Japan Recognition in 2011. He also was awarded Cosmos International Prize as a member of Scientific Steering Committee of Census of Marine Life in 2011.

Secretary-General, International Seabed Authority

Michael W. LODGE

Michael W. Lodge is a British national. He received his LLB from the University of East Anglia, and has an MSc in marine policy from the London School of Economics and Political Science. He is a barrister of Gray's Inn, London. Prior to his election as Secretary-General of the International Seabed Authority in July 2016, he had served as Deputy to the Secretary-General and Legal Counsel. Other professional experiences include serving as Legal Counsel to the ISA (1996-2003); Counsellor to the Round Table on Sustainable Development, OECD (2004-2007); Legal Counsel to the South Pacific Forum Fisheries Agency (1991-1995). He has also held appointments as a Visiting Fellow of Somerville College, Oxford (2012-2013), an Associate Fellow of Chatham House, London (2007) and a member of the World Economic Forum's Global Agenda Council on Oceans (2011-2016).

With 28 years of experience as a public international lawyer, Michael Lodge has a strong background in the field of law of the sea as well as ten years' judicial experience in the UK and South Pacific. He spent many years living and working in the South Pacific and was one of the lead negotiators for the South Pacific Island States of the 1995 UN Fish Stocks Agreement. He has also worked as a consultant on fisheries, environmental and international law in Europe, Asia, Eastern Europe, the South Pacific and Africa.

With extensive knowledge of the United Nations and other international organizations, Michael Lodge has facilitated high-level multilateral and bilateral negotiations at international and regional level. His significant achievements include his pivotal role in the ISA from its inception in 1996 and in helping to create and implement the first international regulatory regime for seabed mining. He also contributed to the future security of global fish stocks by leading the process to create the Western and Central Pacific Fisheries Commission from concept to its establishment as the largest regional fisheries management organization in the world, also serving as the interim executive director of the Commission. He was instrumental in advising the Pew Charitable Trusts on their support for the Global Ocean Commission and also acted as an adviser to the Commission on international law of the sea and ocean policy.

Mr. Lodge has published and lectured extensively on the international law of the sea, with over 25 published books and articles on law of the sea, oceans policy and related issues.

Legal Counsel and Deputy to the Secretary-General, International Seabed Authority

Alfonso ASCENCIO-HERRERA

Alfonso Ascencio-Herrera is the Legal Counsel and Deputy to the Secretary-General of the International Seabed Authority. He received his LL.B in Faculty of Law from National Autonomous University of Mexico, and has an MSc in public international law from King's College, University of London. His professional experiences include serving as Legal adviser, permanent mission of Mexico to the United Nations (2001-2006); Deputy head of mission, embassy of Mexico to the Kingdom of Thailand (2006-2010); Director of international law I (Oceans and the Law of the Sea), Legal adviser's office, ministry of foreign affairs of Mexico (2010-2014); Deputy Head of Mission, Embassy of Mexico to the Republic of Korea (2014-2016).

Director of the Office of Environmental Management and Mineral Resources, International Seabed Authority

Sandor MULSOW

Sandor Mulsow, PhD, MSc, BSc, is the Director of the Office of Environmental Management and Mineral Resources of the International Seabed Authority. He is also tenured He has relevant experience in marine biogeochemistry in sediments, water column, environmental monitoring of sediment water interface from shallow waters to deep sea, deep sea geochemistry and radiochemistry (geochronology) of sediments, nodules, crusts and massive sulphides. He also develops underwater instrumentation for sampling temporal integrated variables (SPI, in situ magnetometry). Currently he is also Professor on Marine Geology at University Austral of Chile (on leave).

Head of Geo-surveys and Geo-resources Sector, Pacific Community (SPC)

Akuila TAWAKE

Akuila Tawake, MSc, is the Head of Geo-surveys and Geo-resources Sector at the Pacific Community (SPC). He is based at the Geoscience Division office in Suva, Fiji. Akuila was also responsible for the European Union funded Pacific Deep Sea Minerals Project that was implemented by SPC in the 15 Pacific ACP States from 2011 to 2016. Prior to his appointment to this post, Akuila was the Senior Adviser – Aggregates Geology at the Pacific Islands Applied Geoscience Commission (SOPAC) serving the Pacific Island Countries in identifying alternative sources of construction materials.

Akuila is a member of the International Marine Minerals Society (IMMS); Society for Mining, Metallurgy and Exploration; and the Fiji Mining and Quarrying Council. Previously, Akuila worked as Exploration / Mine / Project Geologist at an operating mine and two mineral exploration projects in Fiji for seven years. He has relevant experience working with a broad range of stakeholders in the mineral / mining industry including governments, regional and international organisations, academic institutions, donor partners, private sector, civil society organisations, communities and resource owners.

Research Director, Institute of Research and Development (New Caledonia)

Pierre-Yves LE MEUR

Dr Pierre-Yves Le Meur has a PhD in Agricultural sciences and development studies from the Institut National Agronomique Paris-Grignon (1992) and a Habilitation (HDR) in Ethnology and social anthropology from the Ecole des Hautes Etudes en Sciences Sociales (EHESS) (2006). He has worked at the National Universiy of Benin (1992-94), the University of Hohenheim (1994-98), the Johannes Gutenberg University of Mainz (1998-2001 and 2002-03, the GRET (French development NGO, 2003-2007) and at IRD since 2007. He has been working extensively on the politics of belonging and natural resources in West Africa, and since more than 10 years in the Pacific. In New Caledonia where he was based 8 years (2008-2015) he has carried out different research projects on the interplay of local and corporate governance in the mining sector, the land issue and the land-mine nexus, and the interactions between mining, environment and politics. He has run for IRD the expert panel commissioned by the French Polynesian and French Governments to review the prospects of deep-sea mineral resources in French

Polynesia and outline a policy framework in this domain. He is a founding member of the Pacific Centre for Social Responsibility and Natural Resources (PacSen) currently hosted by the Pacific Community (SPC). He is preparing with Colin Filer (ANU Canberra) a book on "Large-scale mines and local-level politics in Papua New Guinea and New Caledonia" (forthcoming, ANU Press).

Senior Adviser Maritime Policy and Governance, IFREMER

Yves HENOCQUE

Yves Henocque is a Maritime Policy and Integrated Coastal and Ocean Management Specialist with over 30 years of research and consultancy in coastal and marine environment. First trained in marine ecology and starting working in the field of aquaculture, he then acquired management and international cooperation skills in Japan, South-East Asian and South Pacific countries as International Cooperation Delegate to the French Research Institute for Sustainable Development of the Sea (IFREMER). At the beginning of the 90s, he settled in the Mediterranean to start a new coastal environmental laboratory within the premises of the IFREMER Mediterranean Center. After a dedicated vocational training in the United States in 1994, he started to practice integrated coastal and ocean management (ICOM) and strategic planning in the Mediterranean and other marine regions, more particularly in the Indian Ocean where he was Project Manager and Team Leader of an Integrated Coastal Management EU-funded project (1995-2000) with the Indian Ocean Commission (IOC) and later in Thailand in the frame of a 5-year coastal habitats and resources management project.

During the last 3 years and till now, he has been working with JAMSTEC as guest researcher setting up new bilateral and multilateral collaborating projects, more particularly in the field of deep sea mineral resources and the crafting of ecosystem-based deep ocean management.

Director of Department of Physical Resources and Deep-Sea Ecosystems, IFREMER

Jean-Marc DANIEL

Jean-Marc Daniel holds a PhD in Earth Sciences from the Pierre et Marie Curie University. For this PhD, he worked on the opening mechanisms of the Tyrrhenian Sea under the supervision of Laurent Jolivet in the geology lab of Ecole Normale Supérieure de Paris headed by X Le Pichon.

He used to work at IFPEN from 1995 to 2015. During these years, his main research topics were the mechanics of natural fracture networks and basin modelling with a special focus on faults hydrodynamics. He was co-director of several PhDs, the last ones being working on the structural geology of the Levant Basin (Eastern Mediterranean Sea). Taking advantage of his expertise, he contributed to the development of one of the leading fracture network simulation software marketed by Beicip-Franlab. He was the head of the structural geology department from 2002 to 2012 before being the deputy director of the Geosciences Division.

He joined Ifremer in 2015. He is now the director of the Physical resources and deep sea ecosystems research. Among geoscience, biology and technological research and innovation devoted to the deep sea, he is heading the team in charge of the exploration of deep sea minerals for France on permits delivered by ISA.

Assistant Director, JAMSTEC

Tomohiko FUKUSHIMA

Dr. Fukushima is working for Research and Development (R&G) Center of Submarine Resources, Japan Agency for Marine-Earth Science and Technology (JAMSTEC). He has graduated the faculty of fisheries, Tokyo University of Fisheries in 1984 (bachelor degree) and graduate school of fisheries science, Tokyo University of Fisheries in 1986 (master degree). In 2004, he has obtained D. Sci. Degree from Kyoto University. The title of doctoral thesis is "Ecological Characteristics of Deep-Sea Benthic Organisms in Relation to Manganese Nodules Development Practices". After awarded the doctoral degree, he continued to research activities in the Ocean Policy Research Foundation (Research Fellow), The University of Tokyo (Associate Professor), and then went to the incumbent.

Head in Center for Environmental Biology and Ecosystem Studies (Biodiversity Resource Conservation Office), National Institute for Environmental Studies (NIES)

Masanobu KAWACHI

Dr. Kawachi is a head of office for Biodiversity Resource Conservation, Center for Environmental Biology and Ecosystem Studies, National Institute for Environmental Studies (NIES), and serves on a leader in project team at NIES for development of new research protocol for sea surface environmental monitoring techniques for marine ecosystem at

seafloor mining site under SIP. He has studied microalgae since graduated student at the University of Tsukuba and completed his Ph D. (1994) focusing on taxonomy and functional roles of haptonema in the Haptophyceae. He worked at Marine Biotechnology Institute in 1994-1998. Since 1998, he is a member of NIES and in charge of the Microbial Culture Collection at NIES. He is also working for the School of Integrative and Global Majors, University of Tsukuba, as a Professor.

Senior Staff Scientist, JAMSTEC

Dhugal J. LINDSAY

Dr. Lindsay's research focuses on mid-water ecology, particularly on gelatinous organisms that are too fragile to be sampled by conventional methods (taxonomy, community structure, biodiversity patterns & relationships to their environment), as well as the development of technology and techniques to describe species & map distributions in space & time, currently focused on determining whether variability in biological systems is natural or anthropogenically-forced. He served on the National Academies of Science (U.S.), Ocean Studies Board, Committee on Future Needs in Deep Submergence Science, the Steering Committee of the Census of Marine Zooplankton (Census of Marine Life: CoML) and is currently Lead for the New Technologies Working Group within the Deep Ocean Stewardship Initiative (DOSI). He received the Compass International Award for outstanding contributions to the advancement of the science and engineering of oceanography and marine technology, Marine Technology Society (U.S.)., in 2009.

Associate Professor, Yokohama City University

Makoto SETA

Dr. Makoto Seta is an Associate Professor of International Law at Yokohama City University, Japan. He holds an LL.M in Public International Law, London School of Economics and Political Science (UK); Ph.D. in Law, Waseda University (Japan); LL.M in Public International Law, Waseda University (Japan); LL.B. in International Course, Waseda University (Japan). He worked as a Research Associate in the Institute of Comparative Law at Waseda University from April 2013 to March 2015. His main interest is the law of the sea, especially ocean governance and universal jurisdiction over maritime crimes. In 2013, his article "Regulation for Private Maritime Security Companies and Its Challenges" received an award by Yamagata Maritime Institute. He published International Law for the Ocean Governance in 2015.

Engineer in SIP Project team for Development of New-Generation Research Protocol for Submarine Resources, JAMSTEC

Miyuki NISHIJIMA

Dr. Miyuki Nishijima is a microbiologist belonging to the Ecosystem Observation and Evaluation Methodology Research Unit, Project Team for Development of New-generation Research Protocol for Submarine Resources, in Japan Agency for Marine-Earth Science and Technology (JAMSTEC). She graduated from Agricultural Science Department, Graduate School of Agriculture, Tokyo University of Agriculture in 1987. She worked at Marine Biotechnology Institute in 1990-2002. From 2002, she jointed Techno Suruga Laboratory Co., Ltd. She is in charge of microbial diversity in the deep sea sediment. She has been on loan to JAMSTEC since 2015.

Researcher in SIP Project team for Development of New-Generation Research Protocol for Submarine Resources, JAMSTEC

Tomo KITAHASHI

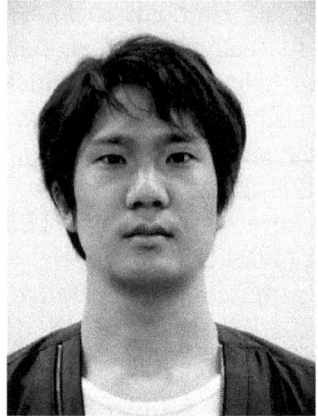

Dr. Tomo Kitahashi is a meiobenthologist belonging to the Ecosystem Observation and Evaluation Methodology Research Unit, Project Team for Development of New-generation Research Protocol for Submarine Resources, in Japan Agency for Marine-Earth Science and Technology (JAMSTEC). He studied in the Graduate school of Frontier Sciences, the University of Tokyo (UT), and completed the research for his Ph.D. (2013) focusing on relationships between meiofaunal diversity and environments. He studied as a postdoctoral fellow at the Department of Marine Ecosystem Dynamics, Atmosphere and Ocean Research Institute (AORI), UT in 2013–2015 and joined JAMSTEC as a researcher in 2015.

Head in Center for Environmental Biology and Ecosystem Studies (Marine Environment Section), National Institute for Environmental Studies (NIES)

Hiroshi KOSHIKAWA

Dr. Koshikawa is working for Center for Regional Environmental Research of National Institute for Environmental Studies (NIES), Japan, as Head of Marine Environmental Section. His research background is in phytoplankton ecology, and he has focused on anthropogenic alterations of marine environments and their impacts on the algal primary production and food webs such as eutrophication, oil pollution, and climate change issues. Since 2014, he has worked on the development of the EIA protocol especially for the marine surface environment at the seafloor mining site where there are risks of heavy metal pollution. He was born in 1968 and obtained the Doctor of Engineering from the Science University of Tokyo in 1997.

Unit Leader in SIP Project team for Development of New-Generation Research Protocol for Submarine Resources, JAMSTEC

Hiroyuki YAMAMOTO

Dr. Hiroyuki Yamamoto is a lead scientist of Environmental Impact Assessment Research Group, R&D Center for Submarine Resources, Japan Agency for Marine-Earth Science and Technology (JAMSTEC), and serves on a unit leader in Project team for Development of

New-Generation Research Protocol for Submarine Resources, Cross-Ministerial Strategic Innovation Promotion Program (SIP) by the Cabinet Office of Japan. He studied in the microbiology laboratory, Faculty of Fisheries Science, Hokkaido University, and completed the research for his Ph.D. (1983) focusing on marine microbial community and interrelation among microbes. After the postdoctoral fellow, he worked as Assistant Professor of Gifu University (1985-1997) and Associate Professor of St. Marianna University (1997-2002). Since 2002, he is a member of JAMSTEC.

Senior Engineer, JAMSTEC

Yasuo FURUSHIMA

Dr. Yasuo Furushima is studying as Senior Engineer in Research and Development (R&D) Center of Submarine Resources Environmental Impact Assessment Research Group, Japan Agency for Marine-Earth Science and Technology (JAMSTEC). He graduated School of Marine Science and Technology, Tokai University in 1988 (bachelor's degree) and Graduate School of Oceanography, Tokai University in 1990 (master degree). In 1996, he got doctor of agriculture in Graduate School of Agricultural and Life Sciences, the University of Tokyo. He was employed as research scientist in JAMSTEC in 1997. His specialty is fisheries oceanology and physical oceanography. He is studying dynamics of bottom mixed layer using an expendable vertical microstructure profiler.

Participants

No.	Name	Organisation
1	Michael Lodge	ISA
2	Alfonso Ascencio-Herrera	ISA
3	Sandor Mulsow	ISA
4	Akuila Tawake	South Pacific Community(SPC)
5	Pierre-Yves Le Meur	Research Director at the Institute of Research and Development
6	Yves Henocque	IFREMER
7	Jean-Marc Daniel	IFREMER
8	Cédric Guillerme	Embassy of France in Tokyo
9	Thibaut Dutruel	Embassy of France in Tokyo
10	Tetsuro Urabe	Cabinet Office, Japan
11	Tomohiko Fukushima	JAMSTEC
12	Koichi Yoshida	Yokohama National University
13	Makoto Seta	Yokohama City University
14	Hiroshi Koshikawa	National Institute for Environmental Studies
15	Tatsuya Nakata	Tokyo University of Marine Science and Technology
16	Hiroyuki Yamamoto	JAMSTEC
17	Yasuo Furushima	JAMSTEC
18	Seiichi Suzuki	Ministry of Education, Culture, Sports, Science and Technology, Japan
19	Takeshi Maki	Ministry of Education, Culture, Sports, Science and Technology, Japan
20	Yasuhiro Kitaura	Ministry of Foreign Affairs, Japan
21	Ikuo Shoji	Ministry of Foreign Affairs, Japan
22	Keisuke Kawamura	Ministry of Foreign Affairs, Japan
23	Eiichi Kikawa	JAMSTEC
24	Kenji Takeda	JAMSTEC
25	Masafumi Shimitashiro	JAMSTEC

No.	Name	Organisation
26	Aki Tanaka	JAMSTEC
27	Yukie Shibata	Yokohama National University
28	Yu Matsunaga	JAMSTEC
29	Akiko Okamatsu	Hosei University
30	Shin Tani	Toyo Construction Co., Ltd.
31	Hiroyuki Nakahara	Research Institute for Ocean Economics
32	Masanobu Kawachi	National Institute for Environmental Studies
33	Takahiro Yamagishi	National Institute for Environmental Studies
34	Syun Tsuboi	National Institute for Environmental Studies
35	Hayuyo Yamaguchi	National Institute for Environmental Studies
36	Shigekatsu Suzuki	National Institute for Environmental Studies
37	Hironori Higashi	National Institute for Environmental Studies
38	Yoshie kasajima	Tokyo University of Marine Science and Technology
39	Nozoji Aoki	Tokyo University of Marine Science and Technology
40	Masahisa Ueda	Japan NUS Co., Ltd.
41	Shuzo Igawa	Japan NUS Co., Ltd.
42	Nozomu Shimizu	Japan NUS Co., Ltd.
43	Yoji Miyata	IDEA Consultants, Inc.
44	Hitoshi Shimoda	Deep Ocean Resources Development
45	Masatsugu Okazaki	Deep Ocean Resources Development
46	Tomoko Tauchi	Deep Ocean Resources Development
47	Kazunori Ishida	KANSO Co., Ltd
48	Yosuke Onishi	KANSO Co., Ltd
49	Mitsuru Shimazu	KANSO Co., Ltd
50	Tanahashi Michio	Kaiyo Engineering Co., Ltd.
51	Hideo Oda	Kaiyo Engineering Co., Ltd.
52	Kentarou Inomata	TechnoSuruga Laboratory Co,. Ltd

No.	Name	Organisation
53	Ryohei Mochizuki	Yokohama National University
54	Asahiko Taira	JAMSTEC
55	Yoshihisa Shirayama	JAMSTEC
56	Hitoshi Hotta	JAMSTEC
57	Toru Sasaki	JAMSTEC
58	Katsuhiko Suzuki	JAMSTEC
59	Dhugal Lindsay	JAMSTEC
60	Tetsuya Miwa	JAMSTEC
61	Tomo Kitahashi	JAMSTEC
62	Hirokazu Itoh	JAMSTEC
63	Masatoshi Nakamura	JAMSTEC
64	Miyuki Nishjijima	JAMSTEC
65	Tatsuo Fukuhara	JAMSTEC
66	Mitsuko Umezu	JAMSTEC
67	Suehiro Nitta	JAMSTEC
68	Emiri Kitami	JAMSTEC
69	Shiro Matsugaura	JAMSTEC
70	Nobuko Unozawa	JAMSTEC
71	Yayoi Yoshino	JAMSTEC
72	Yumi Ueno	JAMSTEC